# WHY DO ANIMALS MOLT AND SHED?

BY ROSIE BANKS

Gareth Stevens PUBLISHING

**Please visit our website, www.garethstevens.com. For a free color catalog of all our high-quality books, call toll free 1-800-542-2595 or fax 1-877-542-2596.**

**Library of Congress Cataloging-in-Publication Data**

Names: Banks, Rosie, 1978- author.
Title: Why do animals molt and shed? / Rosie Banks.
Description: New York : Gareth Stevens Publishing, [2024] | Series: Why do animals do that? | Includes index.
Identifiers: LCCN 2022048600 (print) | LCCN 2022048601 (ebook) | ISBN 9781538285213 (library binding) | ISBN 9781538285206 (paperback) | ISBN 9781538285220 (ebook)
Subjects: LCSH: Molting–Juvenile literature. | Animals–Adaptation–Juvenile literature.
Classification: LCC QL49 .B225 2024 (print) | LCC QL49 (ebook) | DDC 591.4–dc23/eng/20221013
LC record available at https://lccn.loc.gov/2022048600
LC ebook record available at https://lccn.loc.gov/2022048601

Published in 2024 by
**Gareth Stevens Publishing**
2544 Clinton Street
Buffalo, NY 14224



Designer: Claire Wrazin
Editor: Therese Shea

Photo credits: Cover Valmedia/Shutterstock.com; Series Art (texture) Anna Timoshenko; p. 5 Julia Zavalishina/Shutterstock.com; p. 7 Sharon Feragotti/Shutterstock.com; p. 9 Matt Gibson/Shutterstock.com; p. 9 (inset) K Steve Cope/Shutterstock.com; p. 11 Michelle Holihan/Shutterstock.com; p. 11 (inset) moosehenderson/Shutterstock.com; p. 13 (top) David James Chatterton/Shutterstock.com; p. 13 (bottom) godi photo/Shutterstock.com; p. 15 Touched by light images/Shutterstock.com; p. 15 (inset) Skyler Ewing/Shutterstock.com; p. 17 chinahbzyg/Shutterstock.com; p. 19 Buu Photographers/Shutterstock.com.



Printed in the United States of America

CPSIA compliance information: Batch #CSGS24: For further information contact Gareth Stevens, New York, New York at 1-800-542-2595.

# CONTENTS



**Boldface** words appear in the glossary.

## Finding Fur

If you have a furry pet, you might find fur around your home. When an animal loses fur, it's shedding. Shedding is a kind of molting. Molting means getting rid of an outer **layer**. Molting is a helpful **adaptation**!

## Seasonal Shedding

All animals with hair or fur shed. In spring, animals shed to get ready for warm weather. In fall, they shed to prepare to grow a heavier coat for winter. **Chemicals** in the body called hormones tell their body to do this.

bear shedding winter fur

## Shedding Antlers

Male deer grow antlers. They use them to attract, or draw, **mates**. They also fight other deer with their antlers. In winter, the antlers begin to loosen. By spring, the deer sheds its antlers. It grows new, larger antlers.

## What About Horns?

Antlers are all bone. They're shed each year. Horns are bone inside but covered with matter like your fingernails. Horns keep growing throughout an animal's life. They aren't shed. However, **pronghorn** shed the sheath, or covering, of their horns each year.

sheath of a pronghorn's horn

pronghorn

## Molting Feathers

Molting is important to a bird's health. Feathers keep birds warm and dry. They need them for flying too. Birds need to get rid of old or **damaged** feathers. They may shed many feathers at a time or a few.

molting

after molting

When a bird is learning to fly, it's called a fledgling. It has special feathers. It molts these as it becomes an adult. Birds may molt and grow different feathers when it's mating time too. Birds can look odd as they molt!

adult cardinal

fledgling cardinal

## Shedding Outer Shells

Animals called arthropods have a hard covering called an exoskeleton. **Insects**, spiders, and crabs are arthropods. As they grow, they get too big for their exoskeleton. They molt, or shed, their outer covering. A new exoskeleton is under it.

grasshopper shedding exoskeleton

The new exoskeleton is soft at first. It's flexible, or bendable. The animal **stretches** in this exoskeleton. The covering then hardens. The hard shell guards the animal from predators. Young crabs grow quickly. They may molt many times in just a month.

## Amazing Adaptation

Many more kinds of animals molt. When some frogs shed their skin, they get rid of the **fungus** on it. When geckos and other lizards shed their skin, they may eat it. The skin gives their body **nutrients**. Molting is an amazing adaptation!

# WHAT IS MOLTING?

The shedding of

feathers | fur or hair | skin | outer shell | antlers

to make way for new growth.

# GLOSSARY

**adaptation:** A change in a type of animal that makes it better able to live in its surroundings.

**chemical:** Matter that can be mixed with other matter to cause changes.

**damaged:** Harmed or ruined.

**fungus:** A living thing that is somewhat like a plant, but doesn't make its own food, have leaves, or have a green color.

**insect:** A small, often winged, animal with six legs and three main body parts.

**layer:** One thickness of something lying over or under another.

**mate:** One of two animals that come together to make babies.

**nutrient:** Something a living thing needs to grow and stay alive.

**pronghorn:** A horned hoofed animal that lives in the grasslands and deserts of western North America that looks like an antelope.

**stretch:** To extend, or reach out, arms, legs, or other limbs to make the muscles long and tight.

# FOR MORE INFORMATION

## BOOKS

Fields, Hannah. *Keeping Warm with Fur and Fat.* New York, NY: PowerKids Press, 2018.

Klepeis, Alicia. *Snakes Are Awesome.* Mankato, MN: 12 Story Library, 2018.

Mattern, Joanne. *Strange Bodies.* South Egremont, MA: Red Chair Press, 2019.

## WEBSITES

**Why Bugs, Tarantulas, and Other Creatures Shed Their Skin**
*www.nationalgeographic.com/animals/article/why-animals-shed-their-skin*
Learn more about molting from National Geographic.

**Why Do Dogs (and Cats) Shed So Much?**
*animals.howstuffworks.com/pets/dogs-cats-shed.htm*
Read how shedding can be linked to a pet's health.

# INDEX